Ar

by Oswald Boelcke

TRANSLATED FROM THE GERMAN BY ROBERT REYNOLD HIRSCH, M.E.

WITH A FOREWORD BY

JOSEPH E. RIDDER, M.E.

FOREWORD

BY JOSEPH E. RIDDER

An unassuming book, still one of those which grip the reader from beginning to end. When the author started to write his daily impressions and adventures, it was to keep in touch with his people, to quiet those who feared for his safety every moment, and at the same time to give them a clear idea of his life. Without boasting, modestly and naturally, he describes the adventures of an aviator in the great World War. It could well serve as a guide to those who are studying aviation. Although he has avoided the stilted tone of the school-master, still his accomplishments as a knight of the air must fascinate any who know aviation. For the aviators as well as their machines have accomplished wonders. They are rightly called the eyes of the army-- these iron-nerved boys who know no fear. Admiral Schley's historic words after the battle of Santiago: "There will be honor enough for us all" can well be said of the aviators of all nations now at war. For in spite of all enmity the aviators have followed the knightly code of old which respects a good opponent and honors him. Captain Boelcke's death, after his meteoric career, was mourned alike by friend and foe. Great as is the damage done by this war, horrible as is its devastation, it has acted as a tonic on aviation. Before the war, of course, there had been some achievements of note. Since the day when the Wright brothers announced their conquest of the air, man did not rest till the problem was completely solved. And this war, which continually has spurred man to new murderous inventions, has also seen the airplane in action. While at the start of the war the comparatively few airplanes in use were employed as scouts, a few months saw them fitted with machine guns and devices for dropping explosives. Hand in hand with this came the rapid development of the airplane itself. To-day we can truthfully say that a journey, even a long one, by airplane is less dangerous than an automobile ride through a densely populated district. But one thing we must not forget, even though the invention of the airplane by the Wrights is an American one (in spite of the fact that the Wrights give some credit to the German Lilienthal) the Europeans have far outstripped us in the development of this invention.

As sad as it is to say it, we must admit that in regard to aviation America is still in its infancy. Every European nation has outdone us. When, in the summer of 1916, we sent our troops to Mexico, they had only six old machines at their disposal. Instead of relying on these for information, General Pershing had nothing but anxiety for their safety every time they made a flight. But here, too, if all signs are not deceiving, war has helped us to awake. Aside from the activity in our training-schools where thousands of our young men, surpassed by none anywhere, are being trained, the building of our airplanes is taking a great step forward. The experience gained on the other side is helping us here. At first it was the automobile factory that furnished the satisfactory motor. But now through the war the airplane factories have made enormous progress and helped the aviator to attain new marks in speed, reliability and endurance. While this war lasts every improvement in the airplane is utilized to make added destruction. Yet we can not doubt that after the war we will see further progress made in the airplane in the peaceful contests which are to follow.

INTRODUCTION

BY PROF. HERMANN BOELCKE, DESSAU

Oswald Boelcke was born on the 19th of May, 1891, in Giebichenstein, a suburb of Halle on the Saale. Here his father was professor in the high school. His sister, Luise, and his two brothers, Wilhelm and Heinrich, were born before him in Buenos Ayres, Argentina. There his father had had his first position--rector of the German Lutheran School. Later, Oswald's brother Martin was born in Halle and his brother Max in Dessau. Oswald was the first child born to the Boelcke's in Germany. On the 17th of July, the wedding day anniversary of his parents, he was baptized by his uncle, the Rev. Edmund Hartung. This occurred during a vacation spent at his grandmother's, at Freyburg-on-the-Unstrut, in the same church in which his mother had been baptized, confirmed and married, by the same minister. After a year the family moved to Halle, where he could romp joyously on the Viktoria-platz with his two older brothers and his sister.

At the age of four and a half years he moved to Dessau, in 1895, where his father had received a position as professor in the Antoinette School, connected with a teachers' seminary. He had another year and a half of joyous play in this city. Then he was sent to school, and he owed his education to the Friedrichs gymnasium at Dessau, from which he graduated in the Easter of 1911. When he was three years old he had had a severe attack of whooping-cough. This had left a strong tendency to asthma, and was the cause of much trouble at school through illness. In fact, it was a weakness that plagued him with continual colds even to the last few weeks of his life. While still only a youth, he fought this weakness by practising long-distance running, and in 1913 he won second prize in the Army Marathon at Frankfurt. Aside from this, he was perfectly healthy and was always exercising to keep himself so. In his boyhood he learned how to swim while resting on the hands of his father, who was holding him in the waters of the Mulde River. In a few moments, to the amazement of the spectators, he was paddling around in the water like a duck. This is an example of his courage and self-confidence. In the same way he rapidly developed into a skilled, fearless mountain climber under the tuition of his father, when, as a seventeen-year-old boy, he was first taken on such trips. In the Tux district trips were taken from Lauersbach, and the more difficult the climb the more it pleased Oswald. Only when there was real danger was there any joy for him. His mother will never forget the time she witnessed his climbing of the Helenstein. She was on the lower Krieralpe watching. When it was time to descend he, taking huge strides, fairly ran down the slope covered with loose slabs of stone and waited, standing on his head, for his more cautious father and his brother Martin.

His principal, Dr. Wiehmann, said in the words he spoke at Oswald's burial: "He had no mind for books or things studious; in him there burned the desire for action. He was energetic, dynamic, and needed to use his bodily vigor. Rowing, swimming, diving (in which he won prizes as a schoolboy), ball games of all kinds, and gymnastics, he choose as his favorite occupations before he entered his profession as a soldier." He might also have added skating and dancing, for he was a very graceful dancer. His favorite studies were History, Mathematics and Physics. Treitschke's Works and the reports of the General Staff were the books he said he liked best to read. So he was attracted by the military life while still young. Before even his eldest brother thought of it, Oswald wrote him that he yearned to become an officer. In order to fulfil this desire, he decided while still in the third year of school to write to His Majesty the Kaiser that he would like to be an officer, and ask for admission to a cadet school. His parents did not learn of this till his wish was granted, and though putting no obstacles in his path, decided it was better that he finish his schooling before breaking away from "home life." After this, his parents let him join the Telegraphers' Battalion No. 3, at Koblenz, as color guard. They had full confidence in him and his strength of character, and let him leave home with no misgivings. Thanks to his fine physical condition and his enthusiasm, the King's service in the beautiful country of the Rhine and the Moselle was a joy to him. Here he spent many pleasant years, rich in friendship and making ever stronger the family ties. After finishing his schooling as a soldier, he returned to Koblenz from Metz and in the fall was commissioned as a lieutenant.

In this summer he and his brother Martin had the adventure on the Heiterwand, in the Lechtal Alps, which many heard of. He and his brother, in consequence of a heavy fog, lost their way during a difficult climb and after wandering for a day and a night, were rescued by the heroic sacrifices of Romanus Walch, an engineer, and several guides. It was his love for his parents that made him take the way which was impassable except in a few spots, instead of taking the easier south way. On that day, July 26th, his father was to have charge of the opening celebrations at the Anhalt Shelter, situated on the northern face of the Heiterwand. He felt he had to take the shorter, more difficult route so as not to keep his father in suspense on the day of the festivities. Even if he did not spare his parents this anxiety, still he and his brother arrived shortly after the celebrations, in tattered clothes but fresh and shouting in spite of the strain and lack of food.

He wrote with great satisfaction of his work with the telephone division and later with the wireless division. Especially he liked his work in the Taunus, the Odenwald and the Eiffel, with its varying, beautiful scenery which pleased the nature-lover in him. Service with the wireless took him to Darmstadt with a battalion from Koblenz, and it was there that he first came into contact with the aviation corps. They had a school there on the parade grounds. He silently planned to join them, but not till June, 1914, was he able to attain his heart's desire, when he was transferred to the school at Halberstadt. In six weeks his training was completed, and on the day before the mobilization he passed his final examination. On August 1st, on his way to Darmstadt, where he was ordered, he visited his parents in Dessau for an hour. After they had pushed through the throng around the station to a quiet nook inside, he made a confession to them. He had not been in the wireless service at Halberstadt, as they had thought, but had instead been getting his training as an aviator. He had kept this from them so that he should not spoil their vacation in the Alps at Hinter-Tux. This loving care was remembered in this stirring moment and he was forgiven. Still they could not help being frightened at the dangerous work he had chosen; his brother Wilhelm had already joined the aviation corps of the German army as observer. But in the face of the tremendous happenings of those days, personal care and sorrow had to be forgotten. So they parted with him, commending him to the care of God, who rules the air as well as the earth.

Though eager to be off to war, he had to be content with staying in Darmstadt and Trier with the reserves. Finally, on the 1st of September, he was allowed to fly from Trier to the enemy's country. His objective was Sedan. On the way, he landed in Montmedy to visit his brother Wilhelm, who was an observer with the aviation section stationed there. He was ordered to stay there for a time, and had the great satisfaction of being united with his brother, for the division commander ordered him to report to his troop. So the brothers had the good luck to be fighting almost shoulder to shoulder in the Argonnes and the Champagne. If it was possible, they were both in the same machine: Wilhelm as observer, Oswald as pilot. Each knew he could trust the other implicitly. So they were of one heart and one soul in meeting the thousand and one dangers of their daily tasks.

FROM THE BEGINNING OF THE WAR TO THE FIRST VICTORY

HALBERSTADT, AUGUST 1, 1914

Where I will be sent from here, I cannot say as yet. My old mobilization orders commanded me to report to a reconnoitering squadron in the first line, as commander. But these have been countermanded, and I do not know anything about my destination. I expect to get telegraphic orders to-day or to-morrow.

DARMSTADT, AUGUST 3, 1914

Arrived here safe and sound after a slight detour via Cologne. I am very glad that I can spend to-day and to-morrow with B. and my other old friends. Then they go, and only poor I must stay with the Reserve. I think that we will get our turn, too, in two weeks.

TRIER, AUGUST 29, 1914

Arrived here safely. Myself drove a 30 horsepower Opel via Koblenz. Wonderful auto ride!

I managed to get time to pass my third examination in Darmstadt before I left.

F., SEPTEMBER 3, 1914

Started last night with a non-commissioned officer at six o'clock and landed here safely at seven. It was a very pretty flight.

CH., SEPTEMBER 4, 1914

Have been here with the division for two days. As I had no observer along,

Wilhelm has commandeered me. Of course, I like to fly best with Wilhelm, since he has the best judgment and practical experience. As he already knows the country fairly well, he doesn't need a map at all to set his course. We flew over the enemy's positions for about an hour and a half at a height of two thousand eight hundred meters, till Wilhelm had spotted everything. Then we made a quick return. He had found the position of all the enemy's artillery. As a result of his reports, the first shots fired struck home.

When I reached the aviation field the next afternoon two of the planes had already left; Wilhelm also. For me there were written orders to locate the enemy at certain points. At my machine I found the non-commissioned officer who had come with me from Trier; he said he was to go up with me. This seemed odd to me, because I really should have been flying with Wilhelm. I got in and went off with him, since I knew the country from my first flight. We had quite a distance to fly and were under way two and a half hours. I flew over the designated roads that ran through past the Argonne Forest, and with a red pencil marked on the map wherever I saw anything. Above T., at a height of two thousand five hundred meters, we were under heavy fire. I was rather uncomfortable. To the right, below us, we saw little clouds pop up; then a few to the right and left of us. This was the smoke of the bursting artillery shells. Now, I think nothing about such things. They never hit as long as you fly over 2,500 meters high, as we do.

At 7:10 I landed safely here at our camp. And what was the thanks I got for having sailed around over the enemy's lines for over two and a half hours? I got a "call down." I had hardly shut off my engine when Wilhelm came racing over to me. "Where were you? What have you been doing? Are you crazy? You are not to fly without my permission! You're not to go up unless I am along." And more of the same stuff. Only after I had given my word to do as he asked, would he let me alone.

Wednesday evening we had a fine surprise: two of our "missing" returned. They had been forced to land behind the enemy's line because their motor had stopped. They were hardly down when the "Pisangs" (French peasants) came running toward them from every direction. They managed to get into a nearby woods by beating a hasty retreat. Behind them they heard the yelling of the men and women. The woods was surrounded, and they had to hide till night fell. Then they escaped into the Argonne Forest, under cover of darkness although fired on a number of times. Here they spent five days, avoiding French troops. As they had only berries and roots to eat, and could only travel at night, they were almost ready to surrender. But on the morning of the seventh day they heard someone say, in German, "Get on the job, you fool." Those were sweet words to them, for it was a scouting party of German Dragoons. Thus, they got back to us.

M., SEPTEMBER 10, 1914

Yesterday I went along to the light artillery positions, and from there had a good view of the battlefield. There really was nothing to see. There were no large bodies of soldiers, only here and there a rider or a civilian. The only thing you could see was the smoke from bursting shells and the burning villages all about. But if there was nothing to see, there certainly was plenty to hear--the dull noise of the light artillery, the sharp crash of the field pieces and the crackling of small arms. On the way we passed an encampment of reserves. It was a scene exactly like one during the annual manoeuvers; some were cooking, some strolling about, but most of them loafed around on their backs, not paying any attention to the battle at all.

At 5:30 we went up. Now I had a chance to see from the air the same scene I had just beheld from the ground. There was still heavy firing; as far as the eye could see villages were burning. At 7:30 we were down again.

B., SEPTEMBER 16, 1914

Last night three of us tried to take some observations, but all had to come back, as the clouds were too heavy. This morning it was my turn to go up, but it was raining. We have to have the fires going to keep our quarters warm. Next to me a log-fire is burning merrily. My back is baked to a crisp. When my one side gets too hot, I have to turn to give the other a chance to roast. Later some of the telegraphers are coming over and we are going to play "Schafskopf" (a German card game). C'est la guerre!

B., OCTOBER 12, 1914

This evening I received the Iron Cross.

B., OCTOBER 25, 1914

For weeks the weather has been so foggy that we began to consider ourselves as good as retired. But three days ago it began to become bearable again. We took good advantage of it. We were in our machines early in the morning and "worked" till 5:30 at night. I made five flights to-day. First, Wilhelm, as the observer, did some scout work, and later did some range-finding for the artillery. We had agreed that we were to fly above the enemy's positions and then the artillery was to fire. Then it was Wilhelm's duty, as observer, to see where the shells struck and signal to our artillery, with colored lights, if the shots fell short, beyond, to right or left, of the mark. This we do until our gunners find the range. On the 22d, as a result of this, we destroyed one of the enemy's batteries. The next day we wiped out three in three and a half hours. This sort of flying is very trying to observer and pilot alike, as both have to be paying constant attention to business.

Yesterday Wilhelm was at headquarters, and returned with the Iron Cross of the First Class. He has covered a total distance of 6,500 kilometers over the enemy's soil, while I have covered 3,400.

OCTOBER 27, 1914

Wilhelm has discovered nine of the enemy's batteries south of M. and southeast of Rheims, among them being one right next to the cathedral!

NOVEMBER 5, 1914

As the weather is very poor for flights in mid-day, we do most of our flying right after sunrise, about 7:30. Things began to liven up at different points to-day. Our friend, the enemy, had to be taken down a peg, again. Shortly after 7:30 we started. Everything went well, so that we were back in an hour. Then we payed another visit to our artillery. We now fly for four of our batteries, and they only fire when we give them the range. Whenever they have a target, it is destroyed at the first opportunity. So we made two more flights to-day, therefore, a total of three, and put four enemy batteries out of action. We are doing things wholesale now.

NOVEMBER 10, 1914

Wilhelm has now flown a distance of 9,400, I 7,300, kilometers over enemy soil.

LETTER OF NOVEMBER 15, 1914

Mother doesn't need to be afraid that continual flying will affect our nerves. The very opposite is more probable. We get most impatient if we are kept idle a few days because of poor weather. We stand around looking out of the window to see if it isn't clearing up. Nerves can be the excuse for almost anything, I guess.

B., NOVEMBER 30, 1914

I did not get the Fokker as yet. I was to get it at R., Thursday. Too bad. To fly for the artillery, which is our main work just now, the Fokker is very excellent, because of its speed, stability and ease of control. A new machine has been ordered for me at the factory, but I cannot say if I am going to get it, and when.

P., DECEMBER 9, 1914

Bad weather. No important work. Now, we ought to be in the East, where there is something doing.

Yesterday I was in R. and got my Fokker, which had arrived in the meantime. It is a small monoplane, with a French rotary engine in front; it is about half as large as a Taube. This is the last modern machine which I have learned to fly; now I can fly all the types we make in Germany. The Fokker was my big Christmas present. I now have two machines: the large biplane for long flights and the small Fokker for range finding. This 'plane flies wonderfully and is very easy to handle. Now my two children are resting together in a tent, the little one in a hollow, with its tail under the plane of the big one.

P., JANUARY 21, 1915

Since Christmas we have made the following flights: December 24th, an hour and a half; December 25th, one hour; December 30th, one hour; January 6th, one hour; January 12th, four hours; January 18th, two hours. It was poor weather, so we could not do more than this. There isn't much use in flying now, anyhow, as long as we do not want to advance. We are facing each other here for months, and each side knows the other's position exactly. Changes of position, flanking movements, and bringing up of strong reserves, as in open warfare, is a thing of the past when we stick to the trenches, so there is nothing to report. There would be some sense in flying to find the range, but as we do not want to advance at present our artillery does very little firing. It is sufficient at this stage that an airplane takes a peep over the line once in a while, to see if everything is still as they left it.

P., JANUARY 27, 1915

This morning our Captain gave K. and me the Iron Cross of the First Class.

P., APRIL 25, 1915

To-morrow I leave here; I have been transferred to the ---- Flying Squadron, which is just being established. To-morrow I go to Berlin to report at the inspection of aviators.

P., MAY 16, 1915

Safely back in P. The trip was made in comparatively quick time.

P., MAY 17, 1915

We had to leave here this afternoon, after we had hardly arrived. I am very glad. New scenery and something doing.

D., MAY 22, 1915

I had hoped to have plenty to do here, but the weather cancelled our plans. We had plenty of time to establish ourselves, assemble our machines and tune them up with a few flights.

The city is entirely unharmed and the greater part of the inhabitants are still here. The city gives an impression similar to Zerbst--a modern section with cottages and an old section with older houses: the city hall, remains of the old city wall, and so-forth. The inhabitants are prosperous. All the stores, hotels, coffee-houses and café are open. Every day two of my friends (Immelmann and Lieutenant P.) and I go to one of these coffee-houses.

D., MAY 25, 1915

By chance, I witnessed a great military spectacle. As I did not have to fly in the afternoon, I went to the artillery observer's post with our Captain. About four o'clock we reached V.; from here we had another half hour's walk ahead of us. From a distance we could see there was heavy firing going on. The Major, in the company's bomb-proof, told us that the artillery would hardly have time now to avail themselves of airplanes to find the range for them. The French were just at the time trying to get revenge for an attack we made the day before, and the artillery was very busy. From there we went to the observer's post and were very lucky. Our batteries were just firing at the enemy's, our airplanes finding the range for them. Suddenly the non-commissioned officer at the double-periscope yelled over to us that the French were bringing up reinforcements through the communicating trenches. The Lieutenant of Artillery ran over to the field artillery and showed them the beautiful target. Soon after that a few of our shrapnel burst over these positions. Bang! And the enemy was gone. Suddenly a ball of red fire appeared in the first French trench. This meant--shells fall ahead of trenches; place shots further back. Just then, over a front of one and a half kilometers, a whole brigade of Frenchmen rose from the trenches, shoulder to shoulder, a thing I had never seen before. We have to admire them for their courage. In front, the officers about four or five steps in the lead; behind them, in a dense line, the men, partly negroes, whom we could recognize by their baggy trousers. The whole line moved on a run. For the first four hundred meters (in all they had seven hundred meters to cover) we let them come without firing. Then we let them have our first shrapnel. As the artillery knew the exact range, the first shots were effective. Then came the heavier shells. We now opened a murderous fire; it was so loud that we could not hear each other at two paces. Again and again our shells struck the dense masses and tore huge gaps in them, but, in spite of this, the attack continued. The gaps were always quickly closed. Now our infantry took a hand. Our men stood up in the trenches, exposed from the hips up, and fired like madmen. After three or four minutes the attack slackened in spots; that is, parts of the line advanced, others could not. After a quarter of an hour the French on our left wing, which I could see, reached our trenches, shot and stabbed from above, and finally jumped in. Now we could plainly see the hand-to-hand combat: heads bobbing back and forth, guns clubbed (they seemed to be only trying to hit, not kill), glistening bayonets, and a general commotion. On the right wing, things progressed slower, almost at a standstill. In the middle a group jumped forward now and then, and into them the artillery fired with telling effect. We could see men running wildly about, they could not escape our artillery fire. The whole slope was strewn with bodies. After about a quarter of an hour the Frenchmen started to retreat. First one, then two, then three, came out of our trenches, looked all around, and started for their own trenches. In the meantime more troops came up from the rear. But after the first few started to run more came out of the trenches, until finally all were out and retreating. Our men also got out to be able to fire at the retreating enemy to better advantage. Again and again the French officers tried to close up their ranks, rally their men, and lead them anew to the attack.

But in vain, for more and more sought safety in flight. Many dropped--I think more than in the advance. In the center, the French had advanced to within fifty meters of us, and could get no closer. As the retreat started on the left, some in the center also lost heart, and fled like frightened chickens. But almost all were killed. I saw six running away when a shell exploded near them. The smoke disappeared; there were only four left. A second shell, and only one was left. He was probably hit by the infantry. The following proves how completely we repelled their attack: Four Frenchmen rose, waved their arms and ran toward our trench. Two of them carried a severely wounded comrade. Suddenly they dropped their burden and ran faster toward us. Probably their comrades had fired on them. Hardly were these four in our trenches when fifty more of them got up, waved their caps and ran toward us. But the Frenchmen didn't like this, and in a second four well-placed shells burst between them and us; probably they were afraid that there would be a general surrender on the part of their men. The retreat was now general. At 6:15 the main battle was over. Afterward we could see here and there a few Frenchmen running or crawling to their trench.

I was very glad I had the opportunity to see this. From above, we aviators don't see such things.

PILOT OF A BATTLEPLANE

D., JUNE 24, 1915

Yesterday the Crown Prince of Bavaria, our chief, inspected our camp. Here we have gathered samples of about everything that our knowledge of aviation has developed: Two airplane squadrons and one battleplane division. Both airplane squadrons are equipped with the usual biplanes, only we have an improvement: the wireless, by means of which we direct the fire of our artillery. The battleplane squadron is here because there is a lot to do at present on this front (the West). Among them there are some unique machines, for example: a great battleplane with two motors: for three passengers, and equipped with a bomb-dropping apparatus--it is a huge apparatus. Outside of this, there are other battleplanes with machine guns. They are a little larger than the usual run. Then there are some small Fokker monoplanes, also with machine guns. So we have everything the heart can desire. The squadron has only made one flight, but since then the French haven't been over here. I guess something must have proved an eye-opener to them.

JUNE 30, 1915

Rain, almost continuously, since the 22d. I am absolutely sick of this loafing.

Since June 14th, I have a battleplane of my own: a biplane, with 150-horsepower motor. The pilot sits in front; the observer behind him, operating the machine gun, which can be fired to either side and to the rear. As the French are trying to hinder our aerial observation by means of battleplanes, we now have to protect our division while it flies. When the others are doing range-finding, I go up with them, fly about in their vicinity, observe with them and protect them from attack. If a Frenchman wants to attack them, then I make a hawk-like attack on him, while those who are observing go on unhindered in their flight. I chase the Frenchman away by flying toward him and firing at him with the machine gun. It is beautiful to see them run from me; they always do this as quick as possible. In this way, I have chased away over a dozen.

JULY 6, 1915

I succeeded in carrying a battle through to complete victory Sunday morning.

I was ordered to protect Lieutenant P., who was out range-finding, from enemy 'planes. We were just on our way to the front, when I saw a French monoplane, at a greater height, coming toward us. As the higher 'plane has the advantage, we turned away; he didn't see us, but flew on over our lines. We were very glad, because lately the French hate to fly over our lines. When over our ground the enemy cannot escape by volplaning to the earth. As soon as he had passed us we took up the pursuit. Still he flew very rapidly, and it took us half an hour till we caught up with him at V. As it seems, he did not see us till late. Close to V. we started to attack him, I always heading him off. As soon as we were close enough my observer started to pepper him with the machine gun. He defended himself as well as he could, but we were always the aggressor, he having to protect himself. Luckily, we were faster than he, so he could not flee from us by turning. We were higher and faster; he below us and slower, so that he could not escape. By all kinds of manoeuvers he tried to increase the distance between us; without success, for I was always close on him. It was glorious. I always stuck to him so that my observer could fire at close range. We could plainly see everything on our opponent's monoplane, almost every wire, in fact. The average distance between us was a hundred meters; often we were within thirty meters, for at such high speeds you cannot expect success unless you get very close together. The whole fight lasted about twenty or twenty-five minutes. By sharp turns, on the part of our opponent, by jamming of the action on our machine gun, or because of reloading, there were little gaps in the firing, which I used to close in on the enemy. Our superiority showed up more and more; at the end I felt just as if the Frenchman had given up defending himself and lost all hope of escape. Shortly before he fell, he made a motion with his hand, as if to say: let us go; we are conquered; we surrender. But what can you do in such a case, in the air? Then he started to volplane; I followed. My observer fired thirty or forty more shots at him; then suddenly he disappeared. In order not to lose him, I planed down, my machine almost vertical. Suddenly my observer cried, "He is falling; he is falling," and he clapped me on the back joyously. I did not believe it at first, for with these monoplanes it is possible to glide so steeply as to appear to be falling. I looked all over, surprised, but saw nothing. Then I glided to earth and W. told me that the enemy machine had suddenly turned over and fallen straight down into the woods below. We descended to a height of a hundred meters and searched for ten minutes, flying above the woods, but seeing nothing. So we decided to land in a meadow near the woods and search on foot. Soldiers and civilians were running toward the woods from all sides. They said that the French machine had fallen straight down from a great height, turned over twice, and disappeared in the trees. This news was good for us, and it was confirmed by a bicyclist, who had already seen the fallen machine and said both passengers were dead. We hurried to get to the spot. On the way Captain W., of the cavalry, told me that everyone within sight had taken part in the fight, even if only from below. Everyone was very excited, because none knew which was the German and which the French, due to the great height. When we arrived we found officers, doctors and soldiers already there. The machine had fallen from a height of about 1,800 meters. Since both passengers were strapped in, they had not fallen out. The machine had fallen through the trees with tremendous force, both pilot and observer, of course, being dead. The doctors, who examined them at once, could not help them any more. The pilot had seven bullet wounds, the observer three. I am sure both were dead before they fell. We found several important papers and other matter on them. In the afternoon my observer, W., and I flew back to D., after a few rounds of triumph above the village and the fallen airplane. On the following day, the two aviators were buried with full military honors in the cemetery at M. Yesterday we were there. The grave is covered with flowers and at the spot where they fell there is a large red, white and blue bouquet and many other flowers.

I was very glad that my observer, W., got the Iron Cross. He fought excellently; in all, he fired three hundred and eighty shots, and twenty-seven of them hit the enemy airplane.

LETTER OF JULY 16, 1915

... Father asks if it will be all right to publish my report in the newspapers. I don't care much for newspaper publicity, and I do not think that my report is written in a style suitable for newspapers. The people want such a thing written with more poetry and color--gruesome, nerve-wrecking suspense, complete revenge, mountainous clouds, blue, breeze-swept sky--that is what they want. But if the publication of the report will bring you any joy, I will not be against it.

AUGUST 11, 1915

Early August 10th the weather was very poor so that our officer 'phoned in to the city, saying there was no need of my coming out. So I was glad to stay in bed. Suddenly my boy woke me up, saying an English flyer had just passed. I hopped out of bed and ran to the window. But the Englishman was headed for his own lines, so there wasn't any chance of my catching him. I crawled back to bed, angry at being disturbed. I had hardly gotten comfortably warm, when my boy came in again--the Englishman was coming back. Well, I thought if this fellow has so much nerve, I had better get dressed. Unwashed, in my nightshirt, without leggings, hardly half dressed, I rode out to the camp on my motorcycle. I got there in time to see the fellows (not one, but four!) dropping bombs on the aviation field. As I was, I got into my machine and went up after them. But as the English had very speedy machines and headed for home after dropping their bombs, I did not get within range of them. Very sad, I turned back and could not believe my eyes, for there were five more of the enemy paying us a visit. Straight for the first one I headed. I got him at a good angle, and peppered him well, but just when I thought the end was near my machine gun jammed. I was furious. I tried to repair the damage in the air, but in my rage only succeeded in breaking the jammed cartridge in half. There was nothing left to do but land and change the cartridges; while doing this I saw our other monoplanes arrive and was glad that they, at least, would give the Englishmen a good fight. While having the damage repaired, I saw Lieutenant Immelmann make a pretty attack on an Englishman, who tried to fly away. I quickly went up to support Immelmann, but the enemy was gone by the time I got there. In the meantime, Immelmann had forced his opponent to land. He had wounded him, shattering his left arm--Immelmann had had good luck. Two days before I had flown with him in a Fokker; that is, I did the piloting and he was only learning. The day before was the first time he had made a flight alone, and was able to land only after a lot of trouble. He had never taken part in a battle with the enemy, but in spite of that, he had handled himself very well.

AUGUST 23, 1915

On the evening of the 19th I had some more luck.

I fly mostly in the evening to chase the Frenchmen who are out range-finding, and that evening there were a lot of them out. The first one I went for was an English Bristol biplane. He seemed to take me for a Frenchman; he came toward me quite leisurely, a thing our opponents generally don't do. But when he saw me firing at him, he quickly turned. I followed close on him, letting him have all I could give him. I must have hit him or his machine, for he suddenly shut off his engine and disappeared below me. As the fight took place over the enemy's position, he was able to land behind his own lines. According to our artillery, he landed right near his own artillery. That is the second one I am positive I left my mark on; I know I forced him to land. He didn't do it because he was afraid, but because he was hit.

The same evening I attacked two more, and both escaped by volplaning. But I cannot say whether or not I hit them, as both attacks took place over the French lines.

AUGUST 29, 1915

Day before yesterday I flew my Fokker to the division at ----, where from now on I am to serve with the rank of officer. I am to get a newer, more powerful machine--100-horsepower engine. Yesterday I again had a chance to demonstrate my skill as a swimmer. The canal, which passes in front of the Casino, is about 25 meters wide and 2-1/2 meters deep. The tale is told here that there are fish in the water, too, and half the town stands around with their lines in the water. I have never yet seen any of them catch anything. In front of the Casino there is a sort of bank, where they unload the boats. Yesterday, after lunch, I was standing outside the door with T. and saw a French boy climb over the rail, start in fishing and suddenly hop into the water. I ran over to see what he was doing, but he wasn't in sight. This seemed peculiar, so I wasted no time in thought, but dived over after him. This all happened so quickly that T. was just in time to see me go in and did not know what was the matter. I came to the surface, but still alone. Then I saw, not far from me, bubbles and someone struggling in the water. I swam over to him, dived, came up under him, and had him. In the meantime T. and the chauffeur had arrived and T. thought I was going to drown and got ready to go in after me. Finally we got to a nearby boat and T. pulled the boy and me out. When we got to the land the mother of the boy came running up and thanked me most profusely. The rest of the population gave me a real ovation. I must have looked funny, because I had jumped in as I was and the water was streaming off me.

SEPTEMBER 18, 1915

To-day I went to see the boy's parents and they were very grateful. The boy had grown dizzy while standing on the bank and had fallen in. They said they would get the order of the French Legion of Honor for me if they could. That would be a good joke.

Lately, I have flown to the front every evening with Lieutenant Immelmann, to chase the Frenchmen there. As there are usually eight or ten of them, we have plenty to do. Saturday we had the luck to get a French battleplane and between us chase it till it was at a loss what to do. Only by running away did it escape us. The French did not like this at all. The next evening we went out peacefully to hunt the enemy and were struck right away by their great numbers. Suddenly they went crazy and attacked us. They had a new type biplane, very fast, with fuselage. They seemed to be surprised that we let them attack us. We were glad that at last we had an opponent who did not run the first chance he got. After a few vain attacks, they turned and we followed, each of us took one and soon forced them to volplane to earth. As it was already late, we were satisfied and turned to go home. Suddenly I saw two enemy 'planes cruising around over our lines. Since our men in the trenches might think we were afraid, I made a signal for Immelmann to take a few more turns over the lines to show this was not so. But he misunderstood me and attacked one of the Frenchmen, but the latter did not relish this. Meanwhile the second 'plane started for Immelmann, who could not see him, and I naturally had to go to Immelmann's aid. When the second Frenchman saw me coming he turned and made for me. I let him have a few shots so that he turned away when things got too hot for him. That was a big mistake, for it gave me a chance to get him from behind. This is the position from which I prefer to attack. I was close on his heels and not more than fifty meters separated us, so it was not long before I had hit him. I must have mortally wounded the pilot, for suddenly he threw both his arms up and the machine fell straight down. I saw him fall and he turned several times before striking, about 400 meters in front of our lines. Everybody was immensely pleased, and it has been established beyond all doubt that both aviators were killed and the machine wrecked. Immelmann also saw him fall, and was immensely pleased by our success.

M., SEPTEMBER 23, 1915

Sunday night I unexpectedly received a telegram saying I had been transferred. As yet there is no machine here for me, so, for the time being, I have nothing to do.

M., SEPTEMBER 27, 1915

I was casually wandering through the streets; stopped to read the daily bulletins, and there was my name.

It happened the third day of my stay here. As my machines had not yet arrived, the Captain loaned me a Fokker. I was told to be ready at nine o'clock, as the others were to protect the Kaiser, who was breakfasting in a nearby castle. As I wanted to get acquainted with my machine, I went up at a quarter of nine. I was up about three or four minutes when I saw bombs bursting and three or four enemy 'planes flying toward M. I quickly tried to climb to their altitude. This, of course, always takes some time, and by that time the enemy was over M., unloading their bombs on the railroad station. Luckily they hit nothing. After they had all dropped their bombs (there were now ten of them) they turned to go home. I was now about at their altitude, so I started for them. One of the biplanes saw me--it seems they go along to protect the others--and he attacked me from above. Since it is very hard to fire at an opponent who is above you, I let him have a few shots and turned away. That was all the Frenchman wanted, so he turned back. I again attacked the squadron and soon succeeded in getting in range of the lowest of them. I did not fire till I was within a hundred meters, to avoid attracting unnecessary attention. My opponent was frightened and tried to escape. I was right behind him all the while, and kept filling him with well-aimed shots. My only worry was the others, who heard the shots and came to their comrade's rescue. I had to hurry. I noticed I was having some success, because the Frenchman started to glide to earth. Finally, both of us had dropped from 2,500 meters to 1,200. I kept firing at him from behind, as well as I could. In the meantime, however, two of his friends had arrived and sent me several friendly greetings. That isn't very comfortable, and to add to it all, I was without a map above a strange territory and did not know where I was any longer. As my opponent kept flying lower and his companions followed, I had to assume I was behind the enemy's line. Therefore, I ceased my attack and soon, owing to my speed and lack of desire to follow on the part of the French, I left them far behind. Now I had to find my way back. I flew north, and after a time got back to the district around M., which was familiar to me from my days at the officers' school. When I got back I only knew what I have told, and could report only a battle and not a victory. By aid of a map I found I had been over P. ?M. In the afternoon the report came that the infantry on the heights of ---- had seen a biplane "flutter" to earth. The artillery positively reported that the biplane I had fired on had fallen behind the enemy's barbed-wire entanglements. They said the pilot had been dragged to the trenches, dead or severely wounded. Then our artillery had fired at the 'plane and destroyed it. I can only explain the thing this way: I wounded the pilot during the fight; he had tried to glide to earth and land behind his own lines; shortly before landing he lost consciousness or control of his machine; then he "fluttered" to earth; i.e., fell. This was the fourth one.

OCTOBER 17, 1915

Yesterday, the 16th, I shot down a French Voisin biplane near P.

R., NOVEMBER 2, 1915

On the 30th of October we attacked at T. It was our business to break up all scouting on the part of the enemy, and that was difficult that day. The clouds were only 1,500 meters above earth, broken in spots. The French were sailing around behind their front on the 1,400-meter level. Attacked two through the clouds. The first escaped. I got within 100 meters of the second before he saw me. Then he started to run, but that didn't help him any, because I was much faster than he. I fired 500 shots before he fell. Was within three to five meters of him. He would not fall. In the very moment when we seemed about to collide, I turned off to the left. He tilted to the right. I saw nothing more of him. Was very dizzy myself. Was followed by two Farmans and was 1,000 meters behind the enemy's lines. Artillery fired. Too high. Got home without being hit. The enemy airplane fell behind his own lines. The wreck, about 200 meters from our lines, is plainly visible, especially one wing, which is sticking straight up. The attack was rather rash on my part, but on this day of great military value; the French did not come near our position after that.

D., DECEMBER 12, 1915

Am once more in the familiar town of D. Everything is the same as usual. The Captain was very glad that he could give me the life-saving medal. It had just arrived.

D., DECEMBER 31, 1915

Christmas celebrated very nicely and in comfort. Christmas Eve we had a celebration for the men in one of the hangars, which was all decorated. They all received some fine presents. The authorities had sent a package with all kinds of things for each one of them. In the evening we officers also had a little celebration at the Casino; here they also gave out our presents. For me there was a very beautiful silver cup, among other things. This cup was inscribed "To the victor in the air," and was given to me by the Commander-in-Chief of the Aviation Corps. Immelmann received its mate.

Day before yesterday I had a fight with a very keen opponent, who defended himself bravely. I was superior to him and forced him into the defensive. He tried to escape by curving and manoeuvring, and even tried to throw me on the defensive. He did not succeed, but I could not harm him either. All I did accomplish was to force him gradually closer to earth. We had started at 2,800 and soon I had him down to 1,000 meters. We kept whirring and whizzing around each other. As I had already fired on two other enemy craft on this trip, I had only a few cartridges left. This was his salvation. Finally he could not defend himself any more because I had mortally wounded his observer. Now it would have been comparatively safe for me to get him if I had not run out of ammunition at the 800-meter level. Neither of us was able to harm the other. Finally another Fokker (Immelmann) came to my rescue and the fight started all over again. I attacked along with Immelmann to confuse the Englishman. We succeeded in forcing him to within 100 meters of the ground and were expecting him to land any moment. Still he kept flying back and forth like a lunatic. I, by flying straight at him, wanted to put a stop to this, but just then my engine stopped and I had to land. I saw him disappear over a row of trees, armed myself with a flashlight (I had nothing better) and rode over on a horse. I expected that he had landed, but imagine my surprise! He had flown on. I inquired and telephoned, but found out nothing. In the evening the report came that he had passed over our trenches at a height of 100 meters on his way home. Daring of the chap! Not every one would care to imitate him. Immelmann had jammed his gun and had to quit.

JANUARY 8, 1916

On the 5th of January I pursued two Englishmen, overtook them at H.-L. and attacked the first one. The other did not seem to see me; at any rate he kept right on. The fight was comparatively short. I attacked, he defended himself; I hit and he didn't. He had dropped considerably in the meantime, and finally started to sway and landed. I stayed close behind him, so he could not escape. Close to H. he landed; his machine broke apart, the pilot jumped out and collapsed. I quickly landed and found the 'plane already surrounded by people from the nearby village. The Englishmen, whom I interviewed, were both wounded. The pilot, who was only slightly wounded, could talk German; the observer was severely wounded. The former was very sad at his capture; I had hit his controls and shot them to pieces. Yesterday I visited the observer at the hospital; the pilot had been taken away in the meantime. I brought the observer English books and photographs of his machine. He was very pleased. He said he knew my name well.

On the afternoon of the 5th, I made another flight, but everything was quiet. I landed and rode to the city to eat with the rest, because it was getting cloudy again. Just imagine my luck! I was hardly in when a squadron of ten 'planes appeared. I hurried back again and arrived just as they were dropping their bombs on our field. All the helpers were in the bomb-proofs. I howled as if I were being burned alive. At last someone came. I had to take an 80-horsepower machine, because Immelmann, who had remained behind, had already taken my 160-horsepower machine. But with the 80-horsepower machine I could not reach the enemy in time. Then I saw one somewhat separated from the rest. One Fokker had already attacked it, and I went to help him, for I saw I could not overtake the rest. When the Englishman saw both of us on top of him, he judged things were too hot for him, and quickly landed at V., both of us close behind him. The Englishman was alone, still had all his bombs, was unwounded and had only landed through fear.

JANUARY 15, 1916

Now, events come so fast I cannot keep up with them by writing.

On the 11th we had a little gathering that kept me up later than usual, so I did not feel like getting up in the morning. But, as the weather was good, I strolled out to the field and went up about nine o'clock. I flew over to Lille to lie in wait for any hostile aircraft. At first, I had no luck at all. Finally I saw bombs bursting near Ypres. I flew so far I could see the ocean, but am sorry to say I could not find any enemy 'plane. On my way back, I saw two Englishmen, west of Lille, and attacked the nearer one. He did not appreciate the attention, but turned and ran. Just above the trenches I came within gunshot of him. We greeted each other with our machine guns, and he elected to land. I let him go to get at the second of the pair, and spoil his visit, also. Thanks to my good machine, I gradually caught up with him, as he flew toward the east, north of Lille. When I was still four or five hundred meters away from him, he seemed to have seen all he wanted, for he turned to fly west. Then I went for him. I kept behind him till I was near enough. The Englishman seemed to be an old hand at this game, for he let me come on without firing a shot. He didn't shoot until after I started. I flew squarely behind him, and had all the time in the world to aim, because he did not vary a hair from his straight course. He twice reloaded his gun. Suddenly, after only a short while, he fell. I was sure I had hit the pilot. At 800 meters, his machine righted itself, but then dove on, head-foremost, till it landed in a garden in M., northeast of S. The country is very rough there, so I went back to our landing-place and reported by telephone. To my surprise, I heard that at the time Immelmann had shot down an Englishman near P. I had to laugh.

The greatest surprise came in the evening. We were just at dinner when I was called to the 'phone. At the other end was the Commander-In-Chief's Adjutant, who congratulated me for receiving the order Pour le mite. I thought he was joking. But he told me that Immelmann and I had both received this honor at the telegraphic order of the Kaiser. My surprise and joy were great. I went in and said nothing, but sent Captain K. to the 'phone, and he received the news and broke it to all. First, everyone was surprised, then highly pleased. On the same evening I received several messages of congratulation, and the next day, January 13th, had nothing to do all day but receive other such messages.

Everybody seemed elated. One old chap would not let me go, and I didn't escape till I promised to visit him. From all comers I received messages: by telephone and telegraph. The King of Bavaria, who happened to be in Lille with the Bavarian Crown Prince, invited me to dinner for the 14th of January.

Now comes the best of all. On the 14th, that is, yesterday, it was ideal weather for flying. So I went up at nine o'clock to look around. As it was getting cloudy near Lille, I changed my course to take me south of Arras. I was up hardly an hour, when I saw the smoke of bursting bombs near P. I flew in that direction, but the Englishman who was dropping the bombs saw me and started for home. I soon overtook him.

When he saw I intended to attack him, he suddenly turned and attacked me. Now, there started the hardest fight I have as yet been in. The Englishman continually tried to attack me from behind, and I tried to do the same to him. We circled 'round and 'round each other. I had taken my experience of December 28th to heart (that was the time I had used up all my ammunition), so I only fired when I could get my sights on him. In this way, we circled around, I often not firing a shot for several minutes. This merry-go-round was immaterial to me, since we were over our lines. But I watched him, for I felt that sooner or later he would make a dash for home. I noticed that while circling around he continually tried to edge over toward his own lines, which were not far away. I waited my chance, and was able to get at him in real style, shooting his engine to pieces. This I noticed when he glided toward his own lines, leaving a tail of smoke behind him. I had to stop him in his attempt to reach safety, so, in spite of his wrecked motor, I had to attack him again. About 200 meters inside our positions I overtook him, and fired both my guns at him at close range (I no longer needed to save my cartridges). At the moment when I caught up to him, we passed over our trenches and I turned back. I could not determine what had become of him, for I had to save myself now. I flew back, and as I had little fuel left, I landed near the village of F. Here I was received by the Division Staff and was told what had become of the Englishman. To my joy, I learned that, immediately after I had left him, he had come to earth near the English positions. The trenches are only a hundred meters apart at this place. One of the passengers, the pilot, it seems, jumped out and ran to the English trenches. He seems to have escaped, in spite of the fact that our infantry fired at him. Our field artillery quickly opened fire on his machine, and among the first shots one struck it and set it afire. The other aviator, probably the pilot, who was either dead or severely wounded, was burned up with the machine. Nothing but the skeleton of the airplane remains. As my helpers did not come till late, I rode to D. in the Division automobile, because I had to be with the King of Bavaria at 5:30. From D. I went directly on to Lille. King and Crown Prince both conversed with me for quite a while, and they were especially pleased at my most recent success. Once home, I began to see the black side of being a hero. Everyone congratulates you. All ask you questions. I shall soon be forced to carry a printed interrogation sheet with me with answers all filled out. I was particularly pleased by my ninth success, because it followed so close on the Pour le m 開 ite.

S., MARCH 16, 1916

Since March 11th I am here in S. As the lines near Verdun have all been pushed ahead, we were too far in the rear. We saw nothing of the enemy aviators; the reports came too late, so that we were not as timely as formerly. Therefore, they let me pick out a place nearer the lines. I chose a good meadow. I am entirely independent; have an automobile of my own, also a motor truck, and command of a non-commissioned officer and fifteen men. We are so near the front that we can see every enemy airplane that makes a flight in our vicinity. In the first days of our stay here, I had good luck. The weather was good on March 12th. We had a lot to do. I started about eleven to chase two French Farman biplanes, who were circling around over L'homme mort. By the time I arrived there were four of them. I waited for a good chance, and as soon as two of them crossed our front I went for the upper one. There now ensued a pretty little game. The two Frenchmen stuck together like brothers; but I would not let go of the one I had tackled first.

The second Frenchman, on his part, tried to stick behind me. It was a fine game. The one I was attacking twisted and spiralled to escape. I got him from behind and forced him to the 500-meter level. I was very close to him and quite surprised that he had stopped his twisting; but just as I was about to give him the finishing shots, my machine gun stopped. I had pressed down too hard on the trigger mechanism, in the heat of the battle, and this had jammed. The second Frenchman now attacked me, and I escaped while I could. The second fight took place over our lines. The first Frenchman, as I learned later, had gotten his share. He was just able to glide to the French side of the Meuse, and here he landed, according to some reports; others say he fell. I am inclined to believe the former, but probably he could not pick a good spot in which to land, and so broke his machine. From Lieutenant R. I heard that the machine, as well as an automobile, that came to its aid, were set afire by our artillery. I learned further details from Lieutenant B. After landing, one of the aviators ran to the village, returned with a stretcher and helped carry the other one away. Things seem to have happened like this: I wounded the pilot; he was just able to make a landing; then, with the aid of his observer, he was carried off, and our artillery destroyed his machine.

On the following day, the 13th, there was again great aerial activity. Early in the morning I came just in time to see a French battleplane attack a German above Fort Douaumont. I went for the Frenchman and chased him away--it was beautiful to see him go. In the afternoon, I saw a French squadron flying above L'homme mort, toward D. I picked out one of them and went for him. It was a Voisin biplane, that lagged somewhat behind the rest. As I was far above him, I overtook him rapidly and attacked him before he fully realized the situation. As soon as he did, he turned to cross back over the French front. I attacked him strongly, and he tilted to the right and disappeared under me. I thought he was falling; turned to keep him in sight, and, to my surprise, saw that the machine had righted itself. Again I went for him, and saw a very strange sight. The observer had climbed out of his seat and was on the left plane, holding to the struts. He looked frightened, and it was really a sorry plight to be in. He was defenseless, and I hesitated to shoot at him. I had evidently put their controls out of commission, and the machine had fallen.

To right it, the observer had climbed out on the plane and restored its equilibrium. I fired a few more shots at the pilot, when I was attacked by a second Frenchman, coming to the rescue of his comrade. As I had only a few shots left and was above the enemy's line, I turned back. The enemy 'plane glided on a little distance after I left, but finally fell from a low altitude. It is lying in plain sight, in front of our positions east of the village of D.

We have now spoiled the Frenchmen's fun. On March 14th I again attacked one of their battleplanes, and it seemed in a great hurry to get away from me. I accompanied him a little way, playing the music with my machine gun. He descended behind Fort M., as reported later by our soldiers.

MARCH 17, 1916

Last evening I was invited to dine with the Crown Prince. It was very pleasant. He does not value etiquette, and is very unassuming and natural. He pumped all possible information out of me, as he himself admitted later. We had quite a long talk, and on my taking leave he said he would wish for me that I would soon bring down the twelfth enemy.

S., MARCH 21, 1916

Twelve and thirteen followed close on each other. As the weather was fine, we had a lot to do every day. On the 19th I was flying toward D., in the afternoon, to get two Farmans, who were cruising around behind their front. About 12:45 I saw bombs bursting on the west side of the Meuse. I came just in time to see the enemy flying back over his own lines. I thought he had escaped me when I saw him turn and start for one of our biplanes. That was bad for him, because I got the chance to attack him from above. As soon as he saw me, he tried to escape by steep spirals, firing at me at the same time.

But no one who is as frightened as he was ever hits anything. I never fired unless certain of my aim, and so filled him with well-placed shots. I had come quite close to him, when I saw him suddenly upset; one wing broke off, and his machine gradually separated, piece by piece. As there was a south wind, we had drifted over our positions, and he fell into our trenches. Pilot and observer were both killed. I had hit the pilot a number of times, so that death was instantaneous. The infantry sent us various things found in the enemy 'plane, among them a machine gun and an automatic camera. The pictures were developed, and showed our artillery positions.

This morning I started at 9:50, as our anti-aircraft guns were firing at a Farman biplane above ----. The enemy was flying back and forth in the line Ch-- to Ch--. At 10:10 I was above him, as well as another Farman, flying above M. As the Farman again approached our position, I started to attack him. The anti-aircraft guns were also firing, but I imagine they were only finding the range, since their shots did not come near the Frenchman. At the moment when the one Farman turned toward the south, I started for the other, who was flying somewhat lower. He saw me coming, and tried to avoid an engagement by spiral glides. As he flew very cleverly, it was some time before I got within range. At an altitude of five or six hundred meters I opened fire, while he was still trying to reach his own lines. But in pursuing him, I had come within two hundred meters of the road from M. to Ch., so I broke off the attack. My opponent gave his engine gas (I could plainly see the smoke of his exhaust) and flew away toward the southeast. The success I had two hours later reimbursed me for this failure. In the morning, at about eleven o'clock, I saw a German biplane in battle with a Farman west of O. I swooped down on the Farman from behind, while another Fokker came to our aid from above. In the meantime, I had opened fire on the Farman (who had not seen me at all) at a range of eighty meters. As I had come from above, at a steep angle, I had soon overtaken him. In the very moment as I was passing over him he exploded. The cloud of black smoke blew around me. It was no battle at all; he had fallen in the shortest possible time. It was a tremendous spectacle: to see the enemy burst into flames and fall to earth, slowly breaking to pieces.

The reports that I have been wounded in the head, arms, neck, legs, or abdomen, are all foolish. Probably the people who are always inquiring about me, will now discredit such rumors.

APRIL 29, 1916

Thursday morning, at nine, as I arrived in S., after a short trip to Germany, two Frenchmen appeared--the first seen in the last four weeks. I quickly rode out to the field, but came too late. I saw one of our biplanes bring one of the enemies to earth; the other escaped. I flew toward the front at Verdun, and came just in time for a little scrape. Three Frenchmen had crossed over our lines and been attacked by a Fokker, who got into difficulties, and had to retreat. I came to his aid; attacked one of the enemy, and peppered him properly. The whole bunch then took to their heels. But I did not let my friend escape so easily. He twisted and turned, flying with great cleverness. I attacked him three times from the rear, and once diagonally in front. Finally, he spiralled steeply, toppled over and flew for a while with the wheels up. Then he dropped. According to reports from the ---- Reserve Division, he fell in the woods southwest of V., after turning over twice more. That was number 14.

S., MAY 9, 1916

On May 1st I saw an enemy biplane above the "Pfeffereken," as I was standing at our landing station. I started at once, and overtook him at 1,500 meters altitude. It seems he did not see me. I attacked from above and behind, and greeted him with the usual machine-gun fire. He quickly turned and attacked me. But this pleasure did not last long for him. I quickly had him in a bad way, and made short work of him. After a few more twists and turns my fire began to tell, and finally he fell. I then flew home, satisfied that I had accomplished my task. The whole thing only lasted about two minutes.

JUNE 2, 1916

On the 17th of May we had a good day. One of our scout 'planes wanted to take some pictures near Verdun, and I was asked to protect it. I met him above the ---- and flew with him at a great altitude. He worked without being disturbed, and soon turned back without having been fired at. On the way back, I saw bombs bursting at Douaumont and flew over to get a closer view. There were four or five other German biplanes there; I also noticed several French battleplanes at a distance. I kept in the background and watched our opponents. I saw a Nieuport attack one of our machines, so I went for him and I almost felt I had him; but my speed was too great, and I shot past him. He then made off at great speed; I behind him. Several times I was very near him, and fired, but he flew splendidly. I followed him for a little while longer, but he did not appreciate this. Meanwhile, the other French battleplanes had come up, and started firing at me. I flew back over our lines and waited for them there. One, who was much higher than the rest, came and attacked me; we circled around several times and then he flew away. I was so far below him that it was hard to attack him at all. But I could not let him deprive me of the pleasure of following him for a while. During this tilt, I dropped from 4,000 meters to a height of less than 2,000. Our biplanes had also drifted downward.

Suddenly, at an altitude of 4,700 meters, I saw eight of the enemy's Caudrons. I could hardly believe my eyes! They were flying in pairs, as if attached to strings, in perfect line. They each had two engines, and were flying on the line Meuse-Douaumont. It was a shame! Now, I had to climb to their altitude again. So I stayed beneath a pair of them and tried to get at them. But, as they were flying so high and would not come down toward me, I had no success. Shortly before they were over our kite-balloons they turned. So fifteen or twenty minutes passed. Finally I reached their height. I attacked from below, and tried to give them something to remember me by, but they paid no attention to me, and flew home. Just then, I saw two more Caudrons appear, and, thank goodness, they were below me. I flew toward them, but they were already across the Meuse. Just in time, I looked up, and saw a Nieuport and a Caudron coming down toward me. I attacked the more dangerous opponent first, and so flew straight toward the Nieuport. We passed each other firing, but neither of us were hit. I was only striving to protect myself. When flying toward each other, it is very difficult to score a hit because of the combined speed of the two craft. I quickly turned and followed close behind the enemy. Then the other Caudron started to manoeuver the same way, only more poorly than the Nieuport. I followed him, and was just about to open fire when a Fokker came to my aid, and attacked the Caudron. As we were well over the French positions, the latter glided, with the Fokker close behind him. The Nieuport saw this, and came to the aid of his hard-pressed companion; I in turn followed the Nieuport. It was a peculiar position: below, the fleeing Caudron; behind him, the Fokker; behind the Fokker, the Nieuport, and I, last of all, behind the Nieuport. We exchanged shots merrily. Finally the Fokker let the Caudron go, and the Nieuport stopped chasing the Fokker. I fired my last shots at the Nieuport and went home. The whole farce lasted over an hour. We had worked hard, but without visible success. At least, the Fokker (who turned out to be Althaus) and I had dominated the field.

On the 18th of May I got Number 16. Toward evening I went up and found our biplanes everywhere around Verdun. I felt superfluous there, so went off for a little trip. I wanted to have a look at the Champagne district once more, and flew to A. and back. Everywhere there was peace: on earth as well as in the air. I only saw one airplane, in the distance at A. On my way back I had the good luck to see two bombs bursting at M., and soon saw a Caudron near me. The Frenchman had not seen me at all. He was on his way home, and suspected nothing. As he made no move to attack or escape, I kept edging closer without firing. When I was about fifty meters away from them, and could see both passengers plainly, I started a well-aimed fire. He immediately tilted and tried to escape below me, but I was so close to him it was too late. I fired quite calmly. After about 150 shots I saw his left engine smoke fiercely and then burst into flame. The machine turned over, buckled, and burned up. It fell like a plummet into the French second line trenches, and continued to burn there.

On May 20th I again went for a little hunting trip in the Champagne district, and attacked a Farman north of V. I went for him behind his own lines, and he immediately started to land. In spite of this, I followed him, because his was the only enemy machine in sight. I stuck to him and fired, but he would not fall. The pilot of a Farman machine is well protected by the motor, which is behind him. Though you can kill the observer, and riddle the engine and tanks, they are always able to escape by gliding. But in this case, I think I wounded the pilot also, because the machine made the typical lengthwise tilt that shows it is out of control. But as the fight was too far behind the French front, I flew home.

The next day I again had tangible results. In the afternoon I flew on both sides of the Meuse. On the French side two French battleplanes were flying at a great altitude; I could not reach them. I was about to turn back, and was gliding over L'homme mort, when I saw two Caudrons below me, who had escaped my observation till then. I went after them, but they immediately flew off. I followed, and at a distance of 200 meters, attacked the one; at that very instant I saw a Nieuport coming toward me. I was anxious to give him something to remember me by, so I let the Caudrons go and flew due north. The Nieuport came after me, thinking I had not seen him. I kept watching him until he was about 200 meters away. Then I quickly turned my machine and flew toward him. He was frightened by this, turned his machine and flew south. By my attack, I had gained about 100 meters, so that at a range of 100 to 150 meters, I could fill his fuselage with shots. He made work easy for me by flying in a straight line. Besides, I had along ammunition by means of which I could determine the path of my shots. My opponent commenced to get unsteady, but I could not follow him till he fell. Not until evening did I learn from a staff officer that the infantry at L'homme mort had reported the fall of the machine. In the evening, I went out again, without any particular objective, and after a number of false starts I had some success. I was flying north of Bois de ----, when I saw a Frenchman flying about. I made believe I was flying away, and the Frenchman was deceived by my ruse and came after me, over our positions. Now I swooped down on him with tremendous speed (I was much higher than he). He turned, but could not escape me. Close behind the French lines, I caught up with him. He was foolish enough to fly straight ahead, and I pounded him with a continuous stream of well-placed shots. I kept this up till he caught fire. In the midst of this he exploded, collapsed, and fell to earth. As he fell, one wing broke off. So, in one day, I had gotten Numbers 17 and 18.

LEAVE OF ABSENCE

JULY 4, 1916

I was at S. collecting all the equipment of my division. As all the authorities helped me quickly and well, I was ready to move on June 30th. Imagine my bad luck: just on this very day I was destined to make my exit from the stage. It was like this:

Near Verdun there was not much to do in the air. Scouting had been almost dropped. One day, when there was a little more to do than usual, I had gone up twice in the morning and was loafing around on the field. I suddenly heard machine-gun firing in the air and saw a Nieuport attacking one of our biplanes. The German landed and told me, all out of breath:

"The devil is loose on the front. Six Americans are up. I could plainly see the American flag on the fuselage. They were quite bold; came all the way across the front."

I didn't imagine things were quite so bad, and decided to go up and give the Americans a welcome. They were probably expecting it; politeness demanded it. I really met them above the Meuse. They were flying back and forth quite gaily, close together. I flew toward them, and greeted the first one with my machine gun. He seemed to be quite a beginner; at any rate, I had no trouble in getting to within 100 meters of him, and had him well under fire. As he was up in the clouds and flew in a straight course, I was justified in expecting to bring him to earth soon. But luck was not with me. I had just gotten my machine back from the factory, and after firing a few shots my gun jammed. In vain I tried to remedy the trouble. While still bothering with my gun the other "five Americans" were on me. As I could not fire, I preferred to retreat, and the whole swarm were after me. I tried to speed up my departure by tilting my machine to the left and letting it drop. A few hundred meters, and I righted it. But they still followed. I repeated the manoeuver and flew home, little pleased but unharmed. I only saw that the Americans were again flying where I had found them.[A] This angered me and I immediately got into my second machine and went off again. I was hardly 1,500 meters high when with a loud crash my motor broke apart, and I had to land in a meadow at C.

[Footnote A: The result of this was that the English wireless news service asserted the next day: "Yesterday Adjutant Ribie succeeded in bringing down the famous Captain Boelcke in an air battle at Verdun." In the meantime I have relieved him of this misapprehension.]

We made another pretty flight this day. The district around B. and west of Verdun was to be photographed by a scout division. Captain V. was to go over with the squadron, and asked me to go with two other Fokkers to protect them. I went with them, and as I kept close to them, I was right at hand when two French battleplanes attacked. The first one did not approach very close, but the second attacked the biplane which carried Captain V. As he was just then engaged in looking through his binoculars, he did not see the machine approach. The pilot, also, did not notice it till the last moment. Then he made such a sharp turn that Captain V. almost fell out. I came to their aid; the Frenchman started to run. I could hardly aim at him at all, he flew in such sharp curves and zigzags. At 1,800 meters' elevation, I fired a few parting shots and left him. I was sure he would not do us any more harm. As one of the wires to a spark-plug had broken, my engine was not running right, so I turned and went home. The squadron had all the time in the world to take photographs, and was quite satisfied with results. The machine I had attacked was first reported as having fallen, but later this was denied.

Now came the extremely sad news of Immelmann's death. One evening we received word he had fallen. I first thought it was one of the usual rumors, but, to my deep sorrow, it was later confirmed by staff officers. They said his body was being taken to Dresden. I, therefore, immediately asked for leave to fly to D.

It was very impressive. Immelmann lay in the courtyard of a hospital, on a wonderful bier. Everywhere there were pedestals with torches burning on them.

Immelmann lost his life through a foolish accident. Everything the papers write about a battle in the air is nonsense. A part of his propeller broke off and, due to the jerk, the wire braces of the fuselage snapped. The fuselage then broke off. Aside from the great personal loss we have suffered, I feel the moral effect of his death on the enemy is not to be underrated.

I made good use of my chance to again attack the English at D. I liked it so well, I kept postponing my return to S. One evening I flew a Halberstadt biplane; this was the first appearance of these machines at the front. As it is somewhat similar to an English B.-E., I succeeded in completely fooling an Englishman. I got to within fifty meters of him and fired a number of shots at him. But as I was flying quite rapidly, and was not as familiar with the new machine as with the Fokker, I did not succeed in hitting him right away. I passed beneath him, and he turned and started to descend. I followed him, but my cartridge belt jammed and I could not fire. I turned away, and before I had repaired the damage he was gone.

The next day I had two more opportunities to attack Englishmen. The first time, it was a squadron of six Vickers' machines. I started as they were over L., and the other Fokkers from D. went with me. As I had the fastest machine, I was first to reach the enemy. I picked out one and shot at him, with good results; his motor (behind the pilot) puffed out a great quantity of yellow smoke. I thought he would fall any moment, but he escaped by gliding behind his own line. According to the report of our infantry, he was seen to land two kilometers behind the front. I could not finish him entirely, because my left gun had run out of ammunition, and the right one had jammed. In the meantime, the other Fokkers had reached the English. I saw one 160-horsepower machine (Mulzer, pilot,) attack an Englishman in fine style, but as the Englishman soon received aid, I had to come to Mulzer's rescue. So I drove the one away from Mulzer; my enemy did not know I was unable to fire at him. Mulzer saw and recognized me, and again attacked briskly. To my regret, he had only the same success I had had a while before, and as Mulzer turned to go home, I did likewise. In the afternoon, I again had a chance at an Englishman, but he escaped in the clouds.

Meanwhile, the Crown Prince had telephoned once, and our staff officer several times, for me to return. I had at first said I would wait for better weather, so they finally told me to take the train back if it was poor weather. So I saw it was no use, and the next morning I flew back to S. Here I found a telegram for me: "Captain Boelcke is to report at once to the Commander-in-Chief of the Aerial Division. He is to be at the disposal of the Commander-in-Chief of the Army." My joy was great, for I expected to be sent to the Second Army, where the English offensive was just beginning. In the afternoon I reported to the Crown Prince, and there I began to have doubts, for he left me in the dark as to my future. On the next day I reported to the Chief of the Aerial Division at C., and here all my expectations were proven unfounded. For the present, I was not to fly, but was to rest at C. for my "nerves." You can imagine my rage. I was to stay at a watering-place in C. and gaze into the sky. If I had any wish I just needed to express it, only I was not to fly. You can imagine my rage. When I saw that I could do nothing against this decision, I resolved that rather than stay at C. I would go on leave of absence, and at this opportunity see the other fronts. After I telephoned Wilhelm (who was glad rather than sorry for me), my orders were changed to read: "Captain Boelcke is to leave for Turkey and other countries at the request of ----."

Even though this was nothing that replaced my work, it was, at least, a balm for my wounded feelings. I immediately went to S. to pack my things and use the remaining two days to fly as much as possible. I flew twice that night, because I had to utilize the time. In spite of bad weather, I had the luck to meet five Frenchmen the second time I went up. One came within range and I attacked him. He was quite low and above his own trenches, but in my present frame of mind that did not matter to me. I flew toward him, firing both guns, flew over him, turned and started to attack him again, but found him gone. It was very dark by then. When I got home I asked if anyone had seen him fall, but no one knew anything definite.

The next day the weather was bad, and I flew over to Wilhelm to talk over several things and bid him farewell. Picture my surprise, when I read in the afternoon's wireless reports: "Yesterday an enemy machine was brought down near Douaumont." This could only have been my enemy, because, on account of the bad weather, I was the only German who had gone up at that part of the front. I immediately called up the staff officer, and he said yes, it had been a Fokker, yesterday evening, that had brought down the Frenchman, but no one knew who was flying the Fokker. I told him the time, place, and other circumstances, and he seemed very surprised, and forbid me any further flight. He proceeded to make further inquiries. The next morning the further surprising details arrived: The enemy airplane that had been attacked above our first line trenches had fallen in our lines because of heavy south winds. That was very fine for me. Now, my departure from the front was not so bad, because I had brought down another enemy and so had put a stop to any lies the enemy might start about me. The others, my helpers, friends, etc., were well pleased. To put a stop to any more such breaks of discipline, they made me go direct to Ch. It pleased me that I could make four of my mechanics corporals before I left. Three of them got the Iron Cross. In Ch. I had to quickly make my final preparations, get my passes, etc., for my trip, and now I am on the way, Dessau-Berlin. On the day I left I had breakfast with the Kaiser, and he greeted me with:

"Well, well; we have you in leash now."

It is funny that everyone is pleased to see me cooped up for a while. The sorriest part of all is that I am forced to take this leave just at a time when the English offensive is developing unprecedented aerial activity.

VIENNA, JULY 6, 1916

Several incidents happened just before I left Berlin. My train was scheduled to leave the Zoo at 8:06. A half hour before my departure I noticed that my "Pour le melite" was missing. I could not think of leaving without it. I rode to get it; it had been left in my civilian clothes, but my valet had already taken these. Of course, there was no auto in sight, so I had to take a street car, though I was in a hurry. My valet was, in the meantime, packing my things up. The result was that I got to the station just as the train was pulling out. At the same time the valet was at the station at Friedrichstrasse with all the luggage. After riding around a while we met again at our house. Fischer was trembling like a leaf, for he thought it was all his fault. I immediately changed my plan, for the days till the start of the next Balkan train had to be utilized; so I decided on a flight to headquarters in Vienna and Budapest. I had the Aerial Division announce my coming to Vienna, and left that night from the Anhalt Station. As companion, I had a Bohemian Coal Baron, who had only given 30,000,000 marks for war loans; he was very pleasant. Except for a few attacks by autograph collectors, the trip was eventless. In Tetschen, at the border, I was relieved of the bother of customs officials through the kindness of an Austrian officer. It was the lasting grief of my companion that he had to submit to the customs in spite of all the letters of recommendation he had.

JULY 7, 1916

In Vienna I was met by a brother aviator at the station. He took me to the Commander-in-Chief of their Aviation Division, who very kindly gave me a comrade as guide, and placed an auto at my disposal. The same morning I rode to Fischamend. As it was Sunday, I could not do anything in a military way, and so toward evening my guide and I took a trip through Vienna, and I let him point out the spots of interest to me.

JULY 10, 1916

Early in the morning we were on the aviation field at Aspern, which is somewhat like Adlershof. Here I saw some very interesting machines; for the first time I saw an Italian Caproni. Also, I was shown a French machine, in which a crazy Frenchman tried to fly from Nancy to Russia, via Berlin. He almost succeeded. They say he got as far as the east front, and was brought down there after flying almost ten hours. They said he was over Berlin at 12:30 at night. Then there were some very peculiar-looking Austrian 'planes.

In the afternoon I reported to the Colonel, who advised me to see the flying in the mountains near Trient on my way back from the Balkans. I do not know yet whether or not I will be able to do this; it all depends on time and circumstances.

In the late afternoon I went up on the Kahlenberg to see Vienna from there. I took the trip with a man and his wife, whom I had met on the train. They seemed very pleased at having my company, and lost no opportunity to tell me this. To add to my discomfiture, a reporter interviewed me on the way back; he was the first I have met so far. The fellow had heard by chance that I was in Vienna and had followed me for two days. He sat opposite me on the inclined railway and I had a lot of fun keeping him guessing. He was very disappointed that he had no success with me, but finally consoled himself with the thought of having spoken with me. In the evening I strolled around Vienna--the city makes a much quieter impression than Berlin. One feels that Vienna is more a quiet home town than a modern city.

JULY 11, 1916

To avoid the dreary railroad journey from Vienna to Budapest, I am taking the steamer, and will catch the Balkan train at Budapest. In that way I will see and enjoy the scenery much more. Even if the trip cannot compare with one on the Rhine, it is still very beautiful. To Pressburg the country is hilly; then it is flat country, with trees, and often forests, on the banks. On the trip a twelve-year-old boy recognized my face and would not leave me after that. He was a very amusing chap; knew almost the dates of the days on which I had brought down my various opponents. The worst thing he knew of, so he told me, was that his aunt did not even know who Immelmann was. At Komorn the character of the Danube changes completely. The meadows on the right disappear, and hills take their place. The left bank is still rather flat.

From Grau, where I photographed the beautiful St. Johann's Church, to Waitzen, the country resembles the Rhine Valley very much. From Waitzen to Budapest, the country is level, but in the distance one can see wooded hills and the city of Budapest, over which the sun was just setting as we arrived. The most beautiful of all, is Budapest itself. It makes a very imposing impression; to the left, the palace and the old castle; to the right, the hotels and public buildings; above all, the Parliament Building.

JULY 12, 1916

Slept real late and then walked to the castle, where I got a bird's-eye view of the city.

In the afternoon I took a wagon and rode with Lieutenant F. through Ofen to the Margareten Island. We passed the Parliament and went to the city park, where we ate a lot of cake at Kugler's. From there we walked to the docks. The evening, I spent with some Germans.

Budapest makes a very modern impression; some of the women are ultra-modern.

JULY 13, 1916

Slept while passing through Belgrade. Woke up in the middle of Servia, while passing a station where music was playing. Rode along the Morave Valley; it is wide and flanked with hills. There are many cornfields and meadows, with cows grazing. From Nisch (a city of low houses) we passed through a small valley bordered with high, rocky, hills. Along the Bulgarian Morave, Pirot (Bulgaria), the district becomes a plateau, with mountains in the distance. The country is very rocky, and there is very little farming. The nearer you get to Sofia the more the country becomes farm land. Finally, it merges into a broad level plain, with the Balkans in the background. Sofia: a small station, and small houses. It was getting dark.

JULY 14, 1916

Slept through Adrianople on my way to Turkey. Passed through the customs.

Country: Mountainous; little developed; no trees, but now and then villages, with a few little houses, thatched with straw, and scattered. For little stretches the country is covered with bushes. Most of the country is uncultivated, but here and there you see a corn or potato field.

The railroad is a one-track affair, with very few sidings. Service very poor now, due to the war; long waits at the stations.

The people are poorly clothed, with gaudy sashes and queer headpieces. Just at present they are celebrating some fast days.

The women work like the men, but always have a cloth wrapped around their heads. We met a military transport; the men are brown and healthy looking. Their whole equipment seemed German in origin.

Near the ocean, the farming is carried on on a large scale.

At the Bay of Kutshuk, I saw camels grazing, for the first time.

The ocean itself seemed brown, green, violet--all colors. At the shore people were swimming, and there were two anti-aircraft guns mounted.

St. Stefano is an Oriental town in every sense of the word. At the shore there are neat little European houses. Here, there is a wireless station, etc., just as in Johannistal.

Then came Constantinople. From the train, you cannot see much; mostly old, dirty houses, that look as if they were ready to topple over at the first puff of wind.

At the station, I was met by several German aviators, and taken to the hotel.

The evening, I spent with some officers and a number of gentlemen from the German Embassy.

JULY 15, 1916

Early in the morning I rode to the Great Headquarters and reported to Enver Pasha, who personally gave me the Iron Crescent. Enver, who is still young, impressed me as a very agreeable, energetic, man. Then I went through the Bazar, with an interpreter. This is a network of streets, alleys and loopholes, in which everything imaginable is sold. Then went to the Agia Sofia, the largest mosque, and to the Sultan Ahmed, which has been changed to a barracks.

In the afternoon I went to the General (the ship on which the German naval officers live). In the evening we were in the Petit Champ, a little garden in which a German naval band played.

My valet amuses me. He is very unhappy, because he cannot feel at home, and is being cheated right and left by the people. He had pictured Turkey to be an entirely different sort of a place. He was very indignant because the merchants start at three o'clock, at night, to go through the streets selling their wares.

JULY 16, 1916

In the morning I went out to the General with Lieutenant H. to see a U-boat.

In the afternoon, a Greek funeral passed the hotel. The cover of the coffin is carried ahead and the corpse can be seen in the coffin.

Later, I wandered around in Galata and saw the Sultan, who was just coming out of a mosque. First, mounted policemen came; then there was a mounted bodyguard; then adjutant; then the Sultan in a coach with four horses; then the same retinue again, in reverse order.

JULY 17, 1916

This morning, I at last had a chance to see something of their aviation. We rode through the city in an auto: through Stamboul, along the old Byzantine city wall, past the cemetery, and a number of barracks, through the dreary district to St. Stefano, and looked over the aviation station there. Here, Major S. has made himself quite a neat bit out of nothing at all. Naturally, under present conditions, it is very hard for him to get the necessary materials of all sorts.

In the afternoon I was a guest on board the General.

In the afternoon I went with Captain D. and other gentlemen, through the Bosphorus to Therapia, where the German cemetery is wonderfully situated. Then we inspected a shoe factory at Beikos, and, later, went to the Goeben and Breslau, where I had a splendid reception. After a brief inspection of both boats, we ate supper and enjoyed a concert on deck. On leaving, Captain A., commander of the Goeben, drank a toast to me. Who would have believed this possible a few years ago.

JULY 18, 1916

To-day I took a pleasure spin on the Sea of Marmora, with S.'s adjutant, and his motorboat. We passed the Sultan's palace and went to Skutari, where I made a short stop. Then we went to the Princes' Islands, where we landed at Princepu. Princepu is to Constantinople what Grunewald or Wannsee is to Berlin. It is a wonderful island, hilly and situated in the middle of the sea. All the wealthy have summer homes here, and most of Constantinople takes a trip here Saturday and Sunday. In the Casino, from which there is a beautiful view of the sea, we drank coffee. Toward evening we reached home, after first sailing around the neighboring islands, on one of which the captured defender of Kut-el-Amara lives in a very nice villa.

JULY 19, 1916

At nine, we left for Panderma. The Sea of Marmora was quite calm; at first there were some waves, but later it was very still. The ship was filled with natives; quite a few women, and some officers. Panderma: a small seaport (many small sail-boats), situated at the foot of a mountain, and made up, mostly, of small frame houses. We were met by small government vessels, while the others were taken off by native boats. After a short wait, we started our trip in a Pullman car (the train was made up specially for us). As far as Manias the country is monotonous; a few boats on the sea, and quite a few storks. In the Sursulu-Su Valley there are more villages, well-built, meadows, fruit trees, and large herds of oxen and flocks of sheep. A good road runs next to the railroad. Then it became dark. Slept well after a good supper.

JULY 20, 1916

Woke up south of Akbissal. Country very pretty, cultivated and fertile, with many herds of cattle; caravans of camel, with a mule as leader.

The plains became more pretty as we went on. Smyrna is beautifully situated. At the station I met Buddecke and several other men. I got a room in the Hotel Kramer, right at the sea. From my balcony I have a view over the whole Gulf of Smyrna. In the afternoon, I took a walk after reporting to His Excellency Liman-Sanders. Went through the Bazar, which is not so large as in Stamboul.

JULY 21, 1916

At ten we went to the aviation field at Svedi Kos, south of Smyrna. The aviators live in a school. Close to the field there are the tents of a division. The Turkish soldiers made a good impression.

JULY 22, 1916

In the morning went swimming at Cordelio, with several ladies and gentlemen. Buddecke met us with a yacht. We had a fine sail. The view of the hills from the gulf was beautiful.

JULY 23, 1916

In the morning, again went to Cordelio for a swim, and took some jolly pictures.

JULY 24, 1916

Slept late. In the afternoon took a sail with several gentlemen to the future landing spot for seaplanes.

JULY 25, 1916

In the morning I strolled about alone in the outlying parts of Smyrna. Here, things look much more "oriental."

Now I have to take the long trip to Constantinople via Panderma, then to the Dardanelles. I lose eight days this way, for which I am exceedingly sorry. In an airplane, I could make it in two and a half hours, but Buddecke will not let me have any. He has a thousand and one reasons for not giving me one, but I believe he has instructions to that effect.

JULY 29, 1916

On July 28th I went aboard a gunboat bound for Chanak, with a tow. Gallipoli is a village, with a number of outlying barracks. Several houses on the shore were destroyed by gunfire. Arrived in Chanak toward noon, and went to Merten-Pasha to report. In the afternoon I went to the aviation field and flew over Troy--Kum Kale--Sedil Bar, to the old English position. The flight was beautiful, and the islands of Imbros and Tenedos were as if floating on the clear sea. In the Bay of Imbros we could plainly see the English ships. Outside of the usual maze of trenches we could plainly see the old English camps. Close to Thalaka there was an English U-Boat and a Turkish cruiser, both sunk, and lying partly out of water. At Sedil Bar, a number of steamers and a French battleship were aground. The dead, hilly peninsula was plainly visible. At Kilid Bar, there were large Turkish barracks.

JULY 30, 1916

Went on a small steamer to Sedil Bar. We got off a little before we reached our destination, to go over the whole position with a naval officer, who awaited us. The difference between the Turkish and English positions was striking. The English, of course, had had more and better material to work with. Now it is nothing but a deserted wreck. Then I looked at the English landing places. Here, the Englishmen had simply run a few steamers aground to protect themselves. After a hasty breakfast, I flew to D. with M. and from there, along the north shore of the Sea of Marmora, to St. Stefano.

JULY 31, 1916

To-day was Bairam (Turkish Easter). Flags everywhere; people all dressed in their best; large crowds on the street; sale of crescent flowers on the streets, and parades.

AUGUST 1, 1916

After a short stay in the War Department and the Bazar, I left for Constantinople. Enver Pasha travels on the same train. He had me brought to him by his servant at tea time. He was very talkative and interesting, and talked almost only German.

AUGUST 2, 1916

Toward eleven o'clock, after an enjoyable trip through a well-cultivated section of Rumania, I arrived in Sofia, after passing a Turkish military train. Here I was received by a number of German aviators. In the afternoon, took a trip through Sofia, which makes the same impression as one of the central German capitals. Short visit in the cadet school, then went to the large cathedral.

AUGUST 3, 1916

The military finish I noticed in the cadet school the day before impressed me favorably. H. and I went to the aviation field in Sofia; most of the machines were Ottos.

In the afternoon, I went to the flying school with H. Our guide, Captain P., showed us as special attraction a Bl 開 iot, which he had. The school is still in the first stages of development. From there we went to the resort called Banje, which is nicely located.

In the evening, I was at supper with a military attach? and met Prince Kyrill. He interested me very much, and talked quite intelligently about a number of things.

AUGUST 4, 1916

Early in the morning, I reported to the Bulgarian Secretary of War, who conversed with me for a long while. He is small in stature and talks German fluently. Then I visited a cavalry barracks, where I also saw the new machine-gun companies. Toward evening I took a stroll in the Boris Gardens, and admired the beauty of Sofia.

AUGUST 5, 1916

After an audience with the Bulgarian Chief of Staff, I went to Uskub via Kustendil in an auto. Fischer, my valet, who was along, had to get out en route to make all our train arrangements. In Kustendil, I stopped over, and at the Casino I was with the Bulgarian Chief of Staff. Then there was an interesting trip to Uskub, where I arrived at nine o'clock.

AUGUST 6, 1916

In the afternoon I was with General Mackensen, and sat next to him at the table. Mackensen talked with me for quite a while. He is serious-looking, but not nearly as stern as his pictures lead one to believe.

Later, I went by train to Hudova, and reached aviation headquarters, where I was given a fine welcome in the barracks. The aviators all live in wooden shacks, in a dreary neighborhood. This is not an enviable place to be, especially since they have had nothing to do for months.

AUGUST 7, 1916

In the morning I paid a visit to another division of flyers, and with Captain E. I flew up and down the Greek front. Then I went back to Uskub, where I spent the night.

AUGUST 8, 1916

Went back to Sofia in the auto. Had several punctures, which were really funny, because my Bulgarian chauffeur and I could converse by sign language only. On the road, not far from Kumanova, there was a Macedonian fair, which was very interesting. The peasants, in white clothes, danced an odd but pretty dance, to music played on bagpipes and other instruments.

AUGUST 9, 1916

This morning, shortly before I left, I received a Bulgarian medal for courage. This was presented to me by the adjutant of the Minister of War, together with the latter's picture. I am now going to the Austrian headquarters, from where I mean to see the east front. I don't know yet how I will get the time.

AUGUST 10, 1916

In the afternoon, short auto ride; in the evening, reported to General Conrad.

AUGUST 11, 1916

Presented myself at Archduke Frederick's and met General Cramon. At eleven o'clock, went on toward Kovel.

AUGUST 12, 1916

Arrived in Kovel about eight. Reported to General Linsingen.

AUGUST 15, 1916

Rode to Brest, which is gutted by fire.

AUGUST 16, 1916

Reported to General Ludendorff. Before eating was presented to Field Marshal Hindenburg. At table, sat between Hindenburg and Ludendorff. In the afternoon, flew to Warsaw.

AUGUST 17, 1916

Rode to Wilna.

AUGUST 18, 1916

Rode to Kovno and then to Berlin.

TO THE FORTIETH VICTORY (Fleet Battles)

LETTER OF SEPTEMBER 4, 1916

DEAR PARENTS:

To your surprise, you no doubt have read of my twentieth victory. You probably did not expect I would be doing much flying while arranging my new division.

A few days ago two new Fokkers arrived for me, and yesterday I made my first flight. At the front, the enemy was very active. They have grown quite rash. While I was enjoying a peaceful sail behind our lines, one came to attack me. I paid no attention to him (he was higher than I). A little later I saw bombs bursting near P. Here I found a B.-E. biplane, and with him three Vickers' one-man machines, evidently a scout with its protectors. I attacked the B.-E., but in the midst of my work the other three disturbed me so I had to run. One of them thought he could get me in spite of this, and followed me. A little apart from the rest, I offered battle, and soon I had him. I did not let him go; he had no more ammunition left. In descending, he swayed heavily from side to side. As he said later, this was involuntary; I had crippled his machine. He came down northeast of Th. The aviator jumped out of his burning machine and beat about with hands and feet, for he was also afire. I went home to get fresh supplies of cartridges and start anew, for more Englishmen were coming. But I had no success. Yesterday I got the Englishman, whom I had captured, from the prisoners' camp and took him to the Casino for coffee. I showed him our aviation field and learned a lot of interesting things from him. My field is slowly nearing completion and I am exceedingly busy.

SEPTEMBER 17, 1916

In the meantime, I have made my total twenty-five.

Number 21 I tackled single-handed. The fight with this Vickers biplane did not take very long. I attacked him at an angle from behind (the best; to get him from directly behind is not so good, since the motor acts as a protection). In vain he tried to get out of this poor position; I did not give him the chance. I came so close to him that my machine was smutted by the ensuing explosion of his 'plane. He fell, twisting like a boomerang. The observer fell out of the machine before it struck.

Number 22 was quite bold; with his companions, he was sailing over our front, attacking our machines. This was too bad for him as well as one of his friends, who was shot down by two Rumplers. Number 22 fell in exactly the same way as 21 fell the day before, only he landed within his own lines.

Number 23 was a hard one. I had headed off the squadron he was with and picked the second one. He started to get away. The third attacked Lieutenant R., and was soon engaged by Lieutenants B. and R., but, nevertheless, escaped within his own lines. My opponent pretended to fall after the first shots. I knew this trick, and followed him closely. He really was trying to escape to his own lines. He did not succeed. At M. he fell. His wings broke off and the machine broke into pieces. As he lies so far behind our front I did not get a chance to inspect the wreck. Once, however, I flew over it at a very low altitude.

After a short while I saw several Englishmen circling over P. When I got nearer, they wanted to attack me. As I was lower, I paid no attention to them, but turned away. As they saw I would not fight, one of them attacked another German machine. I could not allow this to go on. I attacked him and he soon had to suffer for it. I shot up his gasoline and oil tanks and wounded him in the right thigh. He landed and was captured. That was Number 24.

Number 25 had to wait till the next day. A fleet of seven Englishmen passed over our field. Behind them I rose and cut off their retreat. At P. I got near them. I was the lower and, therefore, almost defenseless. This they took advantage of, and attacked me. Nerve! But I soon turned the tables and got my sights on one of them. I got nice and close to him, and let him have about 500 shots at forty meters. Then he had enough. Lieutenant von R. fired a few more shots at him, but he was finished without them. At H. he fell in a forest and was completely wrecked.

Things are very lively here. The Englishmen always appear in swarms. I regret I did not have enough machines for all my men. Yesterday the first consignment arrived. The other half will come very soon. They shot down two Englishmen yesterday, and there won't be many Englishmen left in a little while.

Yesterday, my officer for special service arrived; he will relieve me of a lot of work. Nevertheless, my time is well occupied, even when not flying. There is a lot to do if one has to make a division out of practically nothing. But it pleases me to see things gradually work out as I plan them.

LATER

In the meantime, things have changed considerably. Two of my men and I got into an English squadron and had a thorough housecleaning. Each of us brought down an Englishman. We are getting along fine; since last night five Englishmen. I shot down the leader, which I recognized by little flags on one of the planes. He landed at E. and set his machine afire. His observer was slightly wounded. When I arrived in an auto they had both been taken away. He had landed because I had shot his engine to pieces.

LETTER OF OCTOBER 8, 1916

Yesterday you read of Number 30, but even that is a back number. Number 31 has followed its predecessors.

On September 17th came Number 27. With some of my men I attacked a squadron of F.-E. biplanes on the way back from C. Of these, we shot down six out of eight. Only two escaped. I picked out the leader, and shot up his engine so he had to land. It landed right near one of our kite-balloons. They were hardly down when the whole airplane was ablaze. It seems they have some means of destroying their machine as soon as it lands. On September 19th six of us got into an English squadron. Below us were the machines with lattice-work tails, and above were some Morans, as protection. One of these I picked out, and sailed after him. For a moment he escaped me, but west of B. I caught up with him. One machine gun jammed, but the other I used with telling effect. At short range, I fired at him till he fell in a big blaze. During all this, he handled himself very clumsily. This was Number 28.

On September 27th I met seven English machines, near B. I had started on a patrol flight with four of my men, and we saw a squadron I first thought was German. When we met southwest of B., I saw they were enemy 'planes. We were lower and I changed my course. The Englishmen passed us, flew over to us, flew around our kite-balloon and then set out for their own front. However, in the meantime, we had reached their height and cut off their retreat. I gave the signal to attack, and a general battle started. I attacked one; got too close; ducked under him and, turning, saw an Englishman fall like a plummet.

As there were enough others left I picked out a new one. He tried to escape, but I followed him. I fired round after round into him. His stamina surprised me. I felt he should have fallen long ago, but he kept going in the same circle. Finally, it got too much for me. I knew he was dead long ago, and by some freak, or due to elastic controls, he did not change his course. I flew quite close to him and saw the pilot lying dead, half out of his seat. To know later which was the 'plane I had shot down (for eventually he must fall), I noted the number--7495. Then I left him and attacked the next one. He escaped, but I left my mark on him. As I passed close under him I saw a great hole I had made in his fuselage. He will probably not forget this day. I had to work like a Trojan.

Number 30 was very simple, I surprised a scout above our front--we call these scouts "H 鍀 chen" (rabbits)--fired at him; he tilted, and disappeared.

The fall of Number 31 was a wonderful sight. We, five men and myself, were amusing ourselves attacking every French or English machine we saw, and firing our guns to test them. This did not please our opponents at all. Suddenly, far below me, I saw one fellow circling about, and I went after him. At close range I fired at him, aiming steadily. He made things easy for me, flying a straight course. I stayed twenty or thirty meters behind him and pounded him till he exploded with a great yellow flare. We cannot call this a fight, because I surprised my opponent.

Everything goes well with me; healthy, good food, good quarters, good companions, and plenty to do.

OCTOBER 19, 1916

My flying has been quite successful in the last few days.

On October 13th some of my men and I got into a fleet of Vickers machines of about equal number. They did not care to fight, and tried to get away. We went after them. I attacked one, saw that Lieutenant K. was already after him, picked another, attacked him above P. and fired two volleys at him. I descended about 400 meters doing this and had to let him go, because two others were after me, which I did not appreciate. He had to land at his artillery positions, however.

On the 15th of October, there was a lot to do. Lately, the English attack at two or three o'clock in the afternoon, because they have the notion that we are asleep. Just at this hour we went out. Between T. and S. we had a housecleaning; that is, we attacked and chased every Englishman we could find. I regret that during this only one fell (M. shot down his fourth). Shortly after that I saw a scout amusing himself above the lines. I attacked and finished him first thing; I guess I must have killed the pilot instantly. The machine crashed to earth so violently that it raised a huge cloud of dust. That was Number 33.

On October 10th, in the afternoon, I got into a fleet of six Vickers' machines. I had a fine time. The English leader came just right for me, and I settled it after the first attack. With the pilot dead, it fell, and I watched till it struck, and then picked out another. My men were having a merry time with the other Englishmen. One Englishman favored me by coming quite close to me, and I followed him close to the ground. Still, by skillful flying, he escaped.

The day was a good one for my command. Lieutenant R. brought down his fifth, and Lieutenant S. got one, so that in all we got five that day.

On the 16th I got Number 35. After some fruitless flying I saw six Vickers over our lines. These I followed, with Lieutenant B. From command--there were also three machines present. Lieutenant Leffers attacked one and forced him to earth (his eighth). The others were all grouped together in a bunch. I picked out the lowest and forced him to earth. The Englishmen did not try to help him, but let me have him, unmolested. After the second volley he caught fire and fell.

It is peculiar that so many of my opponents catch fire. The others, in jest, say it is mental suggestion; they say all I need do is attack one of the enemy and he catches fire or, at least, loses a wing.

The last few days we had poor weather. Nothing to do.

THE LAST REPORTS

OCTOBER 20, 1916

At 10:30 in the morning, five of my men and I attacked a squadron of six F.-E. biplanes, coming from D. The machine I attacked fell in its own lines after first losing its observer.

It is lying, a wreck, five hundred meters west of A.

OCTOBER 22, 1916

11:45--Several of my men and I headed off two enemy biplanes coming from the east. Both fell. The one I attacked was shot apart.

OCTOBER 22, 1916

About 3:40 in the afternoon I saw an English machine attack two of our biplanes. I attacked immediately, and forced him to land, although he tried to escape.

Southwest of the forest at G. he landed in a huge shell-hole and broke his machine. The pilot was thrown out.

OCTOBER 25, 1916

This morning, near M., I brought down an English B.-E. biplane.

OCTOBER 26, 1916

About 4:45 seven of our machines, of which I had charge, attacked some English biplanes west of P.

I attacked one and wounded the observer, so he was unable to fire at me. At the second attack the machine started to smoke. Both pilot and observer seemed dead. It fell into the second line English trenches and burned up. As I was attacked by a Vickers machine after going two or three hundred meters, I did not see this. According to the report of Group A., at A. o. K. 1., a B.-E. machine, attacked by one of our one-man machines, had fallen. This must have been mine.

FROM THE LAST LETTER

... Mother does not need to worry about me; things are not so terrible as she pictures them. She just needs to think of all the experience I have had at this work, not to mention our advantage in knowledge of how to fly and shoot.

Telegram from the front.[B]

"October 28, 1916, 7:30 in the evening.

"Prepare parents: Oswald mortally injured to-day over German lines.
"WILHELM."

[Footnote B: To his sister.]

TRANSCRIBER'S NOTES:

1. Minor changes have been made to correct typesetters' errors; otherwise every effort has been made to remain true to the author's words and intent.

2. In the Introduction, Professor Boelcke quotes a speaker at the funeral service; this quote was left open in error in typesetting; the transcriber has closed the quote where it appears most appropriate.

Made in United States
North Haven, CT
03 March 2024